SAND

Book II

Written and Illustrated by

M. Khalilah Muhammad, Ph.D.

Order this book online at www.trafford.com
or email orders@trafford.com

Most Trafford titles are also available at major online book retailers.

Trafford PUBLISHING® www.trafford.com
North America & international
toll-free: 844 688 6899 (USA & Canada)
fax: 812 355 4082

Our mission is to efficiently provide the world's finest, most comprehensive book publishing service, enabling every author to experience success. To find out how to publish your book, your way, and have it available worldwide, visit us online at www.trafford.com

ISBN: 978-1-6987-1364-9 (sc)

ISBN: 978-1-6987-1363-2 (e)

Library of Congress Control Number: 2022923111

Print information available on the last page.

Trafford rev. 12/08/2022

For my Beautiful Children:

Ashanti
Jameel
Nadirah
Najlah
And
Zafir

Without whom I would never have
been Able to slow down enough
To observe the crystals of
The sand
And
life.

ABOUT THE AUTHOR

Born to an ex-military and retired Firefighter and an Educator, Dr. M. Khalilah Muhammad was always reading and examining books and developing her writing skills and the skill of open expression. Due to her father's military background, Dr. Muhammad traveled extensively. She loves visiting new places and experiencing new horizons.

As a child, Dr. Muhammad would spend time in the yard playing with worms and examining the gardens her parents established. There were floral gardens and vegetable gardens in her yard. She would assist in the upkeep of the gardens and the yard.

After a hard day's work, Dr. Muhammad, would sit on the porch and marvel at the bounties of life. One of her favorite was watching the flight of the leaves. The ability to soar and sail while going through the cycles of life amazed her. The second favorite was taking time to enjoy the beach, waves, and life at rest and reflection.

Dr. Muhammad is an epidemiologist. She studies trends of dis ease and loves traveling and exploring the world. She was told once by a childhood friend to make sure to, "take time and smell the roses". In doing so, she wanted to spend time in the environment and marvel at creation! Dr. Muhammad currently resides in Georgia.

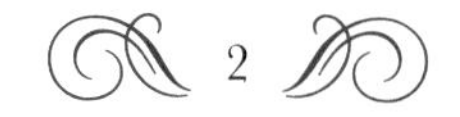

What is Sand?

Where does sand come from?

When it storms, the water from the sea makes waves.

Before the storm, the place we walked on was flat when the sun was out. Now the sky is turning dark.

The birds are not playing. I can chase them! Let us explore on the sand.

The yellow flag is flying high and fast! We must go inside now because the sand looks dark too!

The clouds are gone now and I see the sun.

Mom is running on the beach which is sandy. Maybe she will take us out when she is done.

The sand by the buildings where we play is thick and soft.

We made a castle out of the sand. Sand is silicon dioxide which is natural glass. This is what happens to the rocks that are in the rivers and the streams that are going into the ocean where they are beaten and made even smaller by the waves that we see on the beach that makes the particles tiny and even more fine with age.

We made a moat. Do you know what a moat is?

What will happen when we put water in it?

I like my bucket and shovel. They help me make things in the sand.

Mom gave us a straw from our shakes. We can draw now!

Sand has many colors and textures. The clean sand is where mom takes us to play and explore because it is safer.

My sister drew some words. What do they say?

I AM
@
PEACE

Look! The birds are standing still. This sand is brown and the sand at the other beach was tan.

When the sand is brown like this, it is because of the two minerals: quartz-containing iron oxide and feldspar.

Why does the water push the sand?
What is the white foam?

The bird has one leg! He still does what the other birds do! The foam is still coming on the beach.

Look at the bird claw prints in the sand!

Look at our foot prints in the sun-bleached sand!

The beach here has white sand. It is because of the light colored mineral, quartz.

The beach here has brown sand. Do you remember why it is brown?

There are two types of sand here….why? Do you remember what minerals cause the sand to be brown and tan?

What is pollution?

How can we stop the sand from being dark and polluted?

We walked on the sand and found a few things!

This is the sand in the desert in Dubai, U.A.E. What color is it? Iron oxide in the mineral hematite coats on each individual grain of sand to cause it to have this hue.

There are a lot of Sea Shells in different colors and shapes!

Here in Montego Bay, Jamaica, the sand is tan!

I love the sand in my toes!

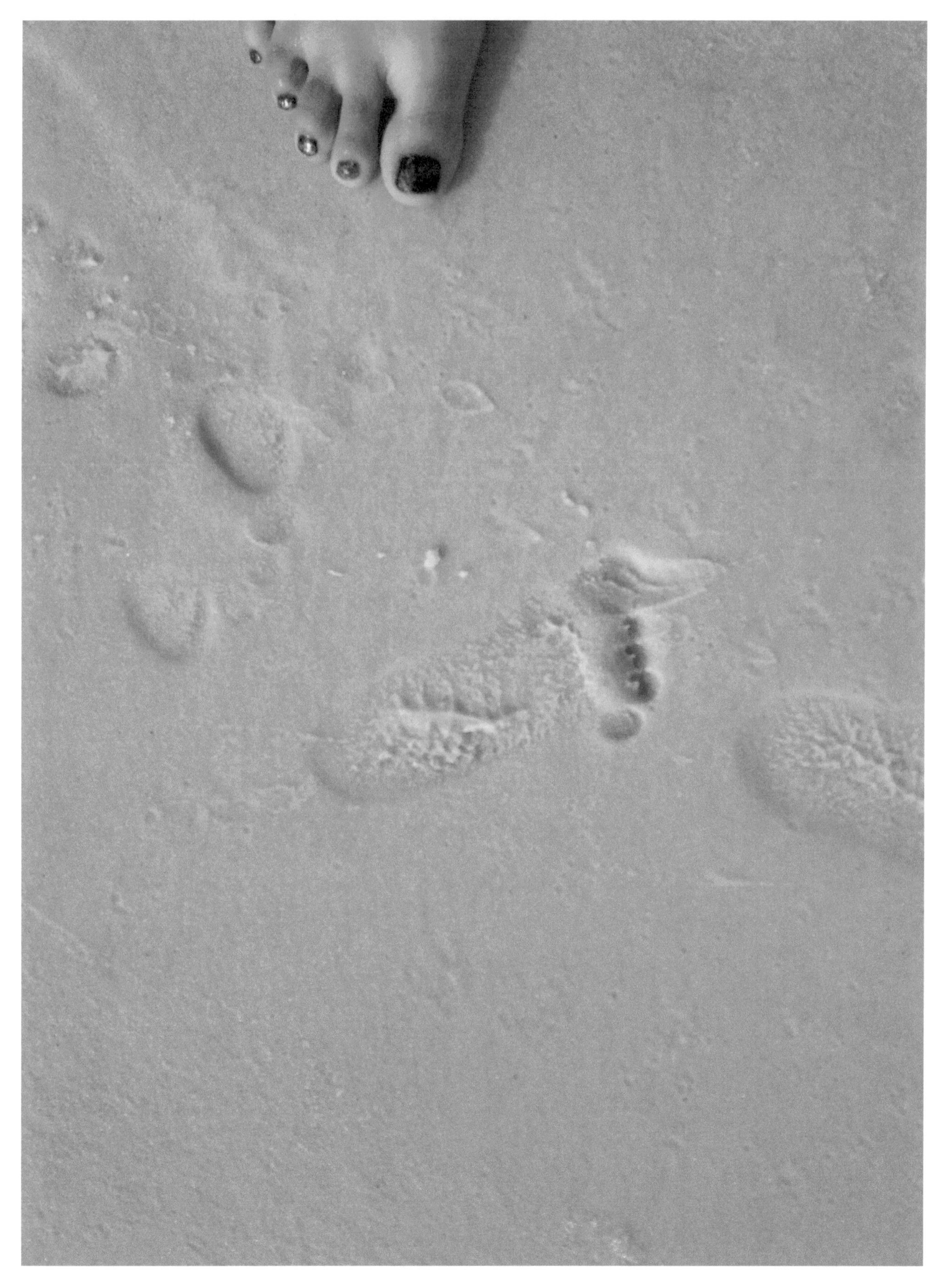

I love the beach which is the place that sand and waves work together for me to be happy!!!!!

Just like here in Jamaica, we have to keep the beach clean by taking our trash with us and clearing out our artwork when we leave.

I am excited that we will go back to the beach again this year! I love the Sand!

WORD FIND

Locate as many words as you can from the list below. Have Fun!

M	R	V	J	O	P	Y	R	V	K	L	M
A	B	U	C	K	E	T	A	U	D	R	C
X	Y	C	W	A	S	R	H	I	N	G	L
J	E	L	L	Y	F	I	S	H	T	O	O
Y	N	A	L	T	E	D	A	F	S	D	U
E	S	W	H	R	E	R	I	B	I	R	D
L	C	P	S	A	N	D	F	O	N	E	S
L	A	R	R	S	E	A	E	E	D	E	H
O	S	I	E	H	D	A	R	K	M	L	O
W	T	N	F	U	C	L	T	O	T	Y	V
H	L	T	S	H	H	R	I	M	W	G	E
E	E	H	H	T	O	S	S	H	E	L	L
V	P	S	R	I	R	C	U	M	T	E	X
H	K	U	I	N	D	A	N	O	F	Y	M
N	G	O	M	R	T	A	W	A	G	V	I
A	D	I	P	O	L	L	U	T	E	D	T

SAND
BUCKET
CLOUD
TRASH
SHRIMP
YELLOW
MOAT
JELLYFISH
BEACH
POLLUTED
WET
SEA
CLAW PRINT
STRAW
DARK
SHELL
SHOVEL
SUN
DIRTY
BIRD
CASTLE

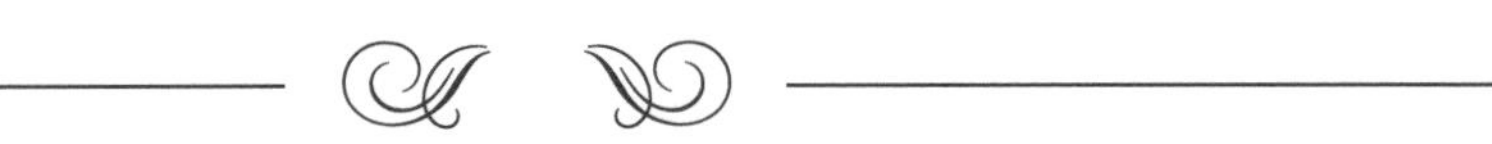

NAME THE BEACHES THAT YOU VISITED AND WHAT YOU FOUND

PICTURE ALBUM OF BEACH VISITS

Printed in the United States
by Baker & Taylor Publisher Services